I0479278

After suffering an attack against his exemplary and impeccable professional career, Engineer Israel Laisequilla offers us in this work a detailed and clearly written guide that allows us to delve into the world of industry in the unique manner that only the so-called "most controversial engineer" can achieve.

INSTRUCTIONS

The information presented here takes into account the access and/or availability of current information and technologies. In case you require more information, formats, and/or examples, your inquiry may increase its reliability thanks to the criteria developed with the book. Any additional information, formulas, and/or videos may be requested using your preferred artificial assistant.

The Bible of Industrial Engineering

ENGR'S WORKSHOP

I. LAISEQUILLA

The bible of industrial engineering

Publishing house IL Editorial

Copyright © 2023 Israel Laisequilla

All rights reserved.

Revision 1

Author's introduction added, July 2024.

Dedicated to anyone who has the courage to dream and the bravery to make their dreams come true.

The bible of industrial engineering

CONTENT

The bible of industrial engineering

ACKNOWLEDGEMENTS

I would like to express my gratitude to all the people who have supported me in the creation of this book:

Firstly, I want to thank my family for their unconditional love and support. Thank you for always being there for me and for understanding my passion for writing.

I also want to thank my friends and colleagues for their support and motivation. Their comments and suggestions were invaluable in completing this project.

I especially appreciate my editor, who gave me a unique opportunity to share my ideas with the world. His guidance and knowledge have been fundamental in bringing this book to its conclusion.

Finally, I want to thank the readers for their interest and dedication. I hope you enjoy this book as much as I enjoyed writing it.

Thanks again to everyone.

engr's Workshop

INTRODUCTION TO INDUSTRIAL ENGINEERING

Industrial engineering is a discipline that focuses on the design, improvement, and management of systems and processes for the production of goods and services. This discipline is based on the principles of engineering, science, and management, and encompasses a wide range of fields including operations management, logistics, production, ergonomics, occupational safety, automation, and control, among others.

The main objective of industrial engineering is to optimize processes and systems to improve efficiency, reduce costs, increase quality, and ensure the safety and well-being of workers and consumers. To achieve these goals, industrial engineers use advanced tools and techniques for analysis, design, and management, as well as a wide variety of technologies and methodologies.

The role of industrial engineering in society is crucial as it enables efficient and effective production of goods and services, which in turn contributes to economic growth and overall well-being. Industrial engineering is used in a wide range of sectors including manufacturing, energy, healthcare, transportation, communications, and financial services.

The evolution of industrial engineering

Industrial engineering has evolved over the years, with its origins dating back to the Industrial Revolution in the 18th century. At that time, the manufacturing industry was in a period of growth and expansion, and the need to increase efficiency and productivity was evident. The invention of the steam engine and other technological advancements allowed for the creation of factories and mass production, which led to a greater demand for engineers and management experts.

Over time, industrial engineering has expanded to encompass a wide range of fields, from production management to logistics and supply chain management. The discipline has also evolved to include greater attention to sustainability and social responsibility as environmental and social issues have become more prominent in society.

The Fundamentals of Industrial Engineering

Industrial engineers use a variety of tools and techniques to optimize processes and systems. Some of these tools include operations research, statistics, simulation, linear programming, optimization, and systems modeling. Industrial engineers also use a variety of management techniques, such as project management, quality management, change management, and supply chain management.

The work of an industrial engineer involves a combination of technical and management skills, which means that industrial engineers must have a strong background in mathematics, physical and social sciences, as well as management and leadership. Additionally, industrial engineers must be able to work in teams, communicate effectively, and have critical thinking skills to address and solve complex problems.

The field of industrial engineering also involves a deep understanding of ergonomics and occupational safety, as industrial engineers work closely with workers in production processes and must ensure that workers are safe and comfortable while performing their tasks.

Another important aspect of industrial engineering is automation and control, which refers to the implementation of systems and technologies to improve efficiency and quality in production processes. Industrial engineers work with robots, control systems, and advanced technologies to optimize processes and reduce costs.

Applications of Industrial Engineering

Industrial engineering is used in a wide variety of industries and sectors. In the manufacturing industry, industrial engineers work on mass production, automation, robotics, and quality improvement. In the healthcare industry, industrial engineers may work on designing and implementing efficient and effective healthcare systems. In the transportation industry, industrial engineers may work on supply chain management and logistics to ensure timely and efficient delivery of goods and services.

In the financial services sector, industrial engineers may work on process and system management to ensure efficiency and security in financial transactions. In the energy sector, industrial engineers may work on improving energy efficiency and implementing sustainable technologies.

In addition, industrial engineering also has applications in the public sector, where industrial engineers may work on improving efficiency and quality in government processes and the provision of public services.

Challenges and Opportunities in Industrial Engineering

Industrial engineering faces a series of challenges and opportunities today. One of the biggest challenges is the need to address environmental and social issues, such as sustainability and social responsibility. Industrial engineers must find ways to reduce the environmental impact of production processes and ensure that business practices are socially responsible.

Another major challenge is the need to adapt to technological advances and the digitalization of production processes. Industrial engineers must stay up-to-date on the latest technologies and tools to ensure that processes and systems are efficient and effective.

In addition, globalization and international competition have increased the need to improve efficiency and quality in production processes. Industrial engineers must find ways to compete in a globalized market and ensure that their processes are efficient and profitable.

Despite these challenges, industrial engineering also offers many exciting opportunities for those seeking a career in this field. The growing demand for industrial engineers means that there is a wide range of employment opportunities in a variety of industries and sectors. In addition, industrial engineering is a constantly evolving discipline, which means that there are many opportunities for professional growth and innovation in the field.

In summary, industrial engineering is a fascinating discipline that combines a wide range of skills and knowledge to improve efficiency and quality in production processes. From supply chain management to quality improvement, automation, and the implementation of advanced technologies, industrial engineers play a critical role in the success of a variety of industries and sectors. With a growing demand for professionals in this field, industrial engineering offers an exciting range of employment and professional growth opportunities.

FUNDAMENTALS OF OPERATIONS MANAGEMENT

Operations management is a key discipline in industrial engineering, as it focuses on the planning, coordination, and control of production and service processes. In this chapter, we will explore the fundamentals of operations management and how they are applied in practice.

Basic concepts of operations management

Operations management focuses on the efficiency and effectiveness of production and service processes. Its goal is to optimize the use of resources, improve the quality of the product or service, and increase customer satisfaction. To achieve this, specific tools and techniques are used, such as production planning, inventory management, production scheduling, quality management, and supply chain management.

Production planning

Production planning is a key process in operations management. It involves defining production goals, determining the necessary resources, and scheduling the activities needed to produce the product or service. Production planning can be divided into three levels: long-term planning, medium-term planning, and short-term planning.

In long-term planning, long-term production goals are established, necessary resources are determined, and investment plans are defined. In medium-term planning, medium-term production plans are defined, activities needed to meet long-term goals are scheduled, and required capacity is determined. In short-term

5

planning, short-term production plans are defined, activities needed to meet medium-term goals are scheduled, and required capacity is determined.

Inventory management

Inventory management refers to the control and monitoring of the quantity and value of products or materials stored in a warehouse. This management is important to ensure the availability of necessary products and materials for production, but also to avoid excessive inventory accumulation that may generate unnecessary costs. Specific techniques such as reorder point, safety inventory, and ABC analysis are used for inventory management.

Production scheduling

Production scheduling is a process that involves the allocation of resources and scheduling of necessary activities to produce the product or service. Specific techniques such as Gantt chart, PERT method, and linear programming are used for production scheduling. These techniques enable effective and efficient production activity scheduling.

Quality management

Quality management refers to a set of activities performed to ensure that a product or service meets customer requirements and expectations. This management is important to ensure customer satisfaction and to avoid unnecessary costs associated with customer dissatisfaction. Specific techniques such as statistical quality control, acceptance sampling, and cause and effect analysis are used for quality management.

Supply chain management

Supply chain management refers to the coordination and management of processes from the acquisition of necessary materials and components for production to the delivery of the final product or service to the end customer. This management is important to ensure the availability of necessary materials and components, as well as to ensure timely and efficient delivery of the product or service. Specific techniques such as demand planning, supplier management, and warehouse management are used for supply chain management.

Tools and Techniques for Operations Management

In addition to the basic concepts of operations management, there are several tools and techniques used to improve the efficiency and effectiveness of production and service processes.

Lean Manufacturing

Lean manufacturing is a methodology that focuses on the elimination of waste in production processes. It is based on five principles: identifying value, mapping the value stream, creating continuous flow, establishing a pull system, and striving for perfection. By implementing lean manufacturing, companies can improve efficiency and reduce costs.

Six Sigma

Six Sigma is a methodology that focuses on reducing variability in production processes. It is based on measuring, analyzing, and improving processes to achieve significant reduction in defects. By implementing Six Sigma, companies can improve the quality of their products or services and reduce costs.

Theory of Constraints

The theory of constraints is a methodology that focuses on identifying and eliminating limitations in production processes. It is based on identifying the most critical constraint and implementing measures to eliminate or reduce it. By implementing the theory of constraints, companies can improve the efficiency of their processes and reduce production times.

Just-in-Time

Just-in-Time is a methodology that focuses on eliminating unnecessary inventory. It is based on the production and delivery of products or materials just when they are needed. By implementing Just-in-Time, companies can reduce the costs associated with inventory management and improve the efficiency of their processes.

Conclusion

In summary, operations management is a key discipline in industrial engineering, focusing on the planning, coordination, and control of production and service processes. Operations management involves production planning, inventory management, production scheduling, quality management, and supply chain management. Additionally, there are specific tools and techniques, such as lean manufacturing, Six Sigma, the theory of constraints, and Just-in-Time, used to improve the efficiency and effectiveness of production and service processes.

DESIGN OF INDUSTRIAL SYSTEMS AND PROCESSES

The design of industrial systems and processes is a fundamental part of industrial engineering. Adequate design of systems and processes can improve efficiency, productivity, quality, and safety in industrial environments. In this chapter, we will explore the fundamentals of industrial systems and processes design, as well as some of the tools and techniques used in this process.

Systems Design

Systems design involves creating a set of interconnected components and subsystems that work together to achieve a specific objective. Systems design can be applied to a wide variety of industrial applications, including manufacturing, transportation, energy, health, and safety.

In system design, it is important to consider the needs and expectations of the end-user, as well as technical and economic constraints. In addition, system design should be able to support future changes and adaptations.

The process of system design involves several stages, which may include defining requirements, identifying design alternatives, evaluating alternatives, selecting the preferred design, and implementing and validating the design.

Process Design

Process design involves creating a set of interconnected activities that convert raw materials and energy into products or services. Proper process design can improve efficiency, quality, safety, and sustainability in industrial environments.

In process design, it is important to consider the needs and expectations of the customer, as well as technical and economic constraints. In addition, process design should be able to support future changes and adaptations.

The process of process design involves several stages, which may include defining requirements, identifying design alternatives, evaluating alternatives, selecting the preferred design, and implementing and validating the design.

Tools and Techniques for Designing Systems and Processes

There are several tools and techniques that can be used in the design of industrial systems and processes. Some of the most common tools and techniques include:

Flowcharts: Flowcharts are tools used to visualize and analyze processes. Flowcharts can help identify problem areas in processes and design solutions to improve the efficiency and effectiveness of processes.

Risk Analysis: Risk analysis is a technique used to identify and evaluate the risks associated with systems and processes. Risk analysis can help companies design solutions to mitigate risks and improve the safety of systems and processes.

Simulation: Simulation is a technique used to model and analyze systems and processes. Simulation can help companies identify problem areas in systems and processes and design solutions to improve the efficiency and effectiveness of systems and processes.

Design for Manufacturability: Design for manufacturability is a technique used for designing industrial systems and processes. In the industry, the design of systems and processes is one of the most important tasks to ensure the quality and efficiency of production. In this chapter, we will explore the importance of designing effective systems and processes in industrial engineering, as well as the key principles that engineers must consider when designing effective systems and processes.

Introduction to Designing Systems and Processes

Designing industrial systems and processes is a key process that involves creating and developing effective systems and processes for the production and manufacturing of products. In general terms, designing industrial systems and processes involves a wide range of activities, from initial planning and design of systems and processes to the implementation and optimization of existing systems and processes.

In industrial engineering, designing systems and processes is a critical task to ensure the quality, efficiency, and profitability of production. Industrial engineers are dedicated to creating optimized systems and processes to maximize productivity, minimize costs, and ensure product quality. By designing effective systems and processes, industrial engineers can improve the competitiveness and profitability of the company.

Key principles of industrial systems and processes design

The design of industrial systems and processes involves a wide range of activities, but there are some key principles that industrial engineers must keep in mind when designing effective systems and processes. These principles include:

Defining the objectives and requirements of the system or process: Before starting the design of a system or process, it is important to clearly define its objectives and requirements. Industrial engineers must take into account the company's objectives, customer needs, and resource limitations when designing a system or process.

Identifying key processes: Effective design of a system or process requires the identification and analysis of the key processes involved. Industrial engineers must understand how processes are performed and how they relate to each other to design effective systems and processes.

Designing for quality: Product quality is a critical objective in the design of industrial systems and processes. Industrial engineers must design systems and processes that minimize variability and maximize product quality.

Designing for efficiency: Efficiency is another important objective in the design of industrial systems and processes. Industrial engineers must design systems and processes that minimize waste of materials, energy, and time.

Designing for flexibility: Industrial systems and processes must be designed to be flexible and adaptable to changes in customer needs or production. Industrial engineers must anticipate possible changes and design systems and processes that can adapt to them.

Continuously evaluating and improving: The design of industrial systems and processes is a continuous process that requires constant evaluation and improvement. Industrial engineers must regularly analyze existing systems and processes to identify opportunities for improvement. In addition to process flow analysis, another tool used in the design of industrial systems and processes is simulation. Simulation allows industrial engineers to create a mathematical model of a system or process and use it to predict its behavior under different conditions. This is especially useful when dealing with complex systems or processes, as it can be difficult to predict how they will perform in real life.

Simulation is often performed using specialized software that can model complex systems and processes. These models can be very detailed and include multiple variables and factors. Industrial engineers can use simulation to test different scenarios and see how they affect system or process performance. They can also use simulation to identify bottlenecks and other performance problems that may not be evident otherwise.

The design of systems and processes also involves selecting appropriate equipment and technologies. Industrial engineers must understand the capabilities and limitations of different types of equipment and technologies in order to select the most suitable one for each task. This may include selecting machinery, tools, materials, and other resources.

Once the appropriate equipment and technology have been selected, industrial engineers can begin designing production processes. This includes creating detailed flowcharts that show the movement of materials and workers through the process. These flowcharts can also help industrial engineers identify areas where efficiency can be improved and costs reduced.

In addition to planning and design, the process of designing industrial systems and processes also includes implementation and commissioning. During this phase, industrial engineers work with plant workers and managers to implement the new systems and processes. They may also provide training and support to ensure that all employees understand the new system or process and are comfortable using it.

Finally, industrial engineers must monitor and measure the performance of the new system or process to ensure that it is functioning as intended. This involves measuring and analyzing key metrics such as performance, efficiency, and costs.

Industrial engineers may also use feedback from employees and managers to identify areas where further improvements can be made.

In summary, the design of industrial systems and processes is a fundamental aspect of industrial engineering. It involves the planning, design, implementation, and measurement of the performance of systems and processes used in industrial production. Industrial engineers use a variety of tools and techniques, including process flow analysis, simulation, and equipment and technology selection, to create efficient and cost-effective systems and processes that help companies achieve their production objectives.

QUALITY CONTROL AND CONTINUOUS IMPROVEMENT

Quality control and continuous improvement are two critical concepts in any organization that seeks to improve its processes and offer high-quality products or services to its customers. Quality control is a process that seeks to ensure that products or services meet customer requirements and specifications and are delivered consistently. Continuous improvement, on the other hand, is a process that seeks to continuously improve the organization's processes and products over time. In this chapter, we will explore the concepts of quality control and continuous improvement in depth, analyzing their key components and their impact on the organization.

Quality Control

Quality control is a process that focuses on ensuring that products or services offered meet customer requirements and specifications. This process is based on identifying problems or deviations in processes and products, implementing measures to correct these problems, and measuring the effectiveness of these measures. Quality control focuses on ensuring that products or services meet the quality specifications established by the organization, industry, or customers.

Quality control is divided into two main types: internal quality control and external quality control. Internal quality control focuses on ensuring that products or services meet the organization's internal requirements and specifications. External quality control, on the other hand, focuses on ensuring that products or services meet the requirements and specifications of customers or external regulations.

Components of Quality Control

There are several key components in quality control. These include planning, process control, quality evaluation, continuous improvement, and a customer-focused approach.

Planning

Planning is a key component of quality control. In this process, quality objectives are established, critical processes and products are identified, quality specifications are set, and quality control plans are developed. Planning also includes identifying risks and implementing measures to mitigate them.

Process Control

Process control is a key component of quality control. This process focuses on ensuring that the organization's processes are running optimally and meet established quality requirements and specifications. Process control includes identifying critical process points, implementing measures to control these points, and measuring the effectiveness of these measures.

Quality Evaluation

Quality evaluation is another key component of quality control. This process focuses on measuring the quality of the products or services offered. Quality evaluation includes measuring the product or service's conformity to established quality specifications, identifying problems or deviations, and implementing measures to correct these problems.

Continuous Improvement

Continuous improvement is a key component of quality control. This process focuses on identifying areas for improvement in the organization's processes and products and implementing measures to continuously improve these processes and products. Continuous improvement is based on Deming's continuous improvement cycle, which consists of four steps: plan, do, check, and act. In the first step, improvements to be implemented are planned, in the second step, planned improvements are implemented, in the third step, the effectiveness of implemented improvements is checked, and in the fourth step, action is taken based on the results obtained to continue improving the process.

Customer Focus

Customer focus is another key component of quality control. This process focuses on understanding customer needs and expectations and ensuring that the products or services offered meet these needs and expectations. Customer focus includes customer feedback and implementing measures to improve customer satisfaction.

Quality Control Tools

There are several quality control tools used to improve processes and ensure the quality of products or services. Some of the most common tools include:

Pareto Chart: It is a tool used to identify critical problems in a process and prioritize them based on their impact on the quality of the product or service.

Ishikawa Diagram: Also known as a fishbone diagram, it is a tool used to identify the causes of a problem and understand how these causes are related.

Control Charts: These are tools used to monitor the quality of a process over time and detect deviations that may indicate problems in the process.

Statistical Sampling: It is a tool used to measure the quality of a process through the collection and analysis of a sample of products or services.

Continuous Improvement

Continuous improvement is a critical process for any organization that seeks to ensure the quality of its products or services and continuously improve its processes. Continuous improvement focuses on identifying areas for improvement and implementing measures to continuously improve the organization's processes and products.

Deming's Continuous Improvement Cycle is a tool used to implement continuous improvement in an organization. This cycle consists of four steps: plan, do, check, and act.

Plan: In this step, areas for improvement are identified and plans are developed to implement improvements in these areas.

Do: In this step, planned improvements are implemented.

Check: In this step, the effectiveness of the implemented improvements is verified through measurement and analysis of the results.

Act: In this step, actions are taken based on the results obtained, and the process continues to be improved.

Continuous improvement is a process that requires the participation and commitment of all members of the organization. It is important to establish a culture of continuous improvement in the organization and encourage the participation of all members in identifying areas for improvement and implementing improvements.

Benefits of Quality Control and Continuous Improvement

Quality control and continuous improvement have many benefits for an organization. Some of these benefits include:

Improvement of the quality of the products or services offered.

Reduction of costs associated with the production and delivery of products or services.

Improvement of the efficiency and effectiveness of processes.

Improvement of customer satisfaction.

Increased customer loyalty and retention.

Reduction of the product return rate or service cancellations.

Improvement of the organization's reputation.

Increase of the organization's competitiveness.

Improvement of the work environment by involving employees in the process of continuous improvement.

Overall, quality control and continuous improvement are fundamental to the long-term success of an organization. By implementing these processes, an organization can ensure that its products or services meet the expectations of its customers and continuously improve its processes to remain competitive in a constantly changing market.

Application Example of Quality Control and Continuous Improvement

To illustrate how quality control and continuous improvement are applied in practice, the example of a car manufacturing company can be used.

Planning: In this phase, areas for improvement in the automobile manufacturing process are identified. In this case, the painting process can be identified as a possible area for improvement, as some inconsistencies in the color and quality of the paint have been identified in some of the manufactured vehicles.

Doing: In this phase, the planned improvements for the painting process are implemented. Improvements can be made to the paint mixing and application process to ensure that it is applied uniformly and meets the required color and quality standards.

Verification: In this phase, the effectiveness of the implemented improvements is verified through measurement and analysis of the results. In this case, the quality of the paint applied to the manufactured vehicles can be measured and compared with previous results to determine if the improvements have had a positive impact.

Acting: In this phase, action is taken based on the results obtained, and the process continues to be improved. If the results are positive, the improvements in the painting process can be implemented in all future manufactured vehicles. If the results are unsatisfactory, other areas for improvement can be identified, and the continuous improvement process can start again.

Conclusions

In conclusion, quality control and continuous improvement are critical processes for any organization seeking to ensure the quality of its products or services and continually improve its processes. These processes can help reduce costs, increase efficiency and effectiveness, improve customer satisfaction, and increase the organization's competitiveness.

Customer focus is fundamental to the success of quality control and continuous improvement as it allows organizations to understand customer needs and expectations and ensure that their products or services meet these needs and expectations.

Quality control tools are valuable for identifying areas for improvement and ensuring the quality of products or services. However, it is important to remember that these tools should be used in conjunction with a holistic approach to continuous improvement to achieve optimal results.

In summary, quality control and continuous improvement are dynamic and ongoing processes that require the participation and commitment of all members of the organization. By implementing these processes effectively, an organization can ensure the quality of its products or services, meet customer needs and expectations, and remain competitive in a constantly changing market.

METHODS OF ANALYSIS AND OPTIMIZATION

The chapter on methods of analysis and optimization is one of the fundamental pillars in decision-making and problem-solving in a wide variety of fields. In this chapter, techniques and tools are explored for analyzing and optimizing systems, processes, projects, among others. On this occasion, we will address the main concepts and techniques used in the optimization and analysis of systems and processes, as well as some practical examples of their application.

Fundamental concepts

Before delving into the techniques and tools for analyzing and optimizing systems and processes, it is important to define some fundamental concepts. Firstly, optimization refers to the process of searching for the best possible solution within a set of possible options. Secondly, analysis refers to the process of understanding how a system or process works, identifying its strengths and weaknesses, and detecting opportunities for improvement.

Another fundamental concept in the optimization and analysis of systems and processes is that of the objective function. The objective function is a quantitative measure of what is desired to be optimized or maximized. For example, in a production process, the objective function may be to maximize production and minimize costs. In a transportation system, the objective function may be to minimize travel time or maximize fuel efficiency.

Optimization methods

There are numerous methods and techniques for optimizing systems and processes, each with its own advantages and disadvantages. Below are some of the most common methods.

Trial and error method

The trial and error method is one of the simplest methods for optimizing a system or process. In this method, different configurations or adjustments are tested until the best solution is found. Although it is an intuitive and easy-to-implement method, it can be inefficient and time-consuming.

Gradient Descent Method

The Gradient Descent Method is an iterative method for finding the minimum of an objective function. In this method, a random point is selected, and the direction and magnitude of the gradient of the objective function are calculated. Then, the point moves in the opposite direction to the gradient, in which the objective function decreases. This process is repeated until a local minimum is reached.

Random Search Method

The Random Search Method consists of randomly generating a solution and evaluating its objective function value. Then, new random solutions are generated and evaluated, and the process is repeated until the best possible solution is found. Although this method is simple and easy to implement, it can be very inefficient and require many evaluations of the objective function.

Linear Programming Method

Linear programming is a mathematical method for optimizing a linear objective function subject to linear constraints. In this method, decision variables are defined, and constraints are established on them. Then, the optimal solution that maximizes or minimizes the objective function subject to the constraints is sought. Linear programming is a powerful and efficient method for solving linear optimization problems.

Integer Programming Method

Integer programming is an extension of linear programming, in which the decision variables are restricted to integer values. This makes the problem more complex and difficult to solve, but it allows modeling a wide variety of real-world problems, such as resource allocation and production scheduling.

Simulation Method

The simulation method involves creating a mathematical model of the system or process that needs to be optimized, and then simulating its behavior to evaluate different scenarios and configurations. This method allows for analyzing the impact of different decisions on the system or process without having to conduct real-world experiments. Simulation is a powerful tool for the analysis and optimization of complex systems and processes.

Sensitivity Analysis Method

The sensitivity analysis method is used to evaluate how the objective function changes when the model parameters are altered. For example, if the objective function is the production of a factory, the parameters could be the costs of materials, processing times, etc. Sensitivity analysis helps identify which parameters have a significant impact on the objective function and which are less important.

Practical Example

To illustrate the use of some of these optimization and analysis methods, let's consider the following example. A transportation company wants to optimize its truck fleet to minimize fuel costs and maximize transport efficiency. The company has a fleet of trucks of different sizes and capacities and needs to decide which trucks to assign to which routes to minimize costs.

To solve this problem, we can use a linear programming model to optimally assign the trucks to the routes. The model can include decision variables for truck assignment to routes, as well as constraints on truck capacity and route demand. The objective function can be a combination of fuel costs and transport efficiency, weighted by their relative importance.

Once the model has been constructed, a linear programming software can be used to find the optimal solution. Sensitivity analysis can then be performed to evaluate how the optimal solution changes when the model parameters, such as fuel prices or route demand, are altered.

Another approach to optimize the truck fleet is to use the simulation method. In this approach, a mathematical model of the transportation system is constructed and its behavior is simulated to evaluate different scenarios and configurations. For example, the impact of changing the truck assignment to routes or adding new trucks to the fleet can be simulated.

Conclusions

The chapter on Methods of Analysis and Optimization is essential for problem-solving and decision-making in a wide variety of fields. In this chapter, we have reviewed the fundamental concepts of optimization and analysis, as well as the most common methods used in practice. Additionally, we have presented a practical example of the application of these methods in the context of optimizing a fleet of trucks.

It is important to note that although there are different methods of analysis and optimization, it is crucial to select the appropriate method for each specific problem. For example, linear programming is suitable for problems that can be modeled as linear equations, while integer programming is necessary when decision variables must be restricted to integer values. Similarly, the simulation method is suitable for complex systems and processes that cannot be easily modeled using mathematical equations.

Furthermore, it is important to remember that the mathematical models used in optimization and analysis are simplifications of reality, and there will always be limitations and assumptions in the model. Therefore, it is important to validate the results obtained through simulation or sensitivity analysis by comparison with real-world data.

In summary, the chapter on Methods of Analysis and Optimization is fundamental for problem-solving and decision-making in a wide variety of fields. The methods presented in this chapter, such as linear programming, integer programming, the simulation method, and sensitivity analysis, are powerful tools for the analysis and optimization of complex systems and processes. However, it is important to select the appropriate method for each specific problem, validate the results, and consider the limitations and assumptions in the model used.

SIMULATION MODELS AND DECISION-MAKING

Simulation is a widely used tool in decision-making in different fields, from engineering to finance and health. It allows modeling complex systems and predicting their behavior in different situations. Decision-making based on simulation involves the use of mathematical models to simulate different scenarios and evaluate the impact of each one. In this chapter, simulation models and their application in decision-making will be discussed.

Types of Simulation Models

There are several types of simulation models used in different fields. Some of the most common types are as follows:

Discrete event models: This type of model is used to simulate systems in which events occur at specific and discrete times. Examples include queues in supermarkets or airports, production processes, and transportation systems.

Dynamic system models: This type of model is used to simulate systems in which variables change continuously over time. Examples include weather systems, economic systems, and biological systems.

Agent-based simulation models: This type of model is used to simulate systems in which individual agents have autonomous behaviors and can interact with each other. Examples include traffic simulations, market simulations, and animal behavior simulations.

Each type of model has its own advantages and disadvantages, and the choice of model type depends on the system being simulated and the objectives of the simulation.

Stages of Simulation

The simulation process consists of several stages, including problem formulation, model construction, model validation, simulation execution, and results analysis.

Problem Formulation: In this stage, the problem to be simulated is defined and simulation objectives are established. The variables that influence the system are also identified and model parameters are defined.

Model Construction: In this stage, the mathematical model representing the system to be simulated is constructed. The type of model to be used is selected and equations describing the system's behavior are established.

Model Validation: In this stage, it is verified that the model is valid and accurately reflects the behavior of the real system. Simulation results are compared with historical or experimental data to evaluate the accuracy of the model.

Simulation Execution: In this stage, the simulation is executed using the model constructed and parameters defined in the previous stages.

Results Analysis: In this stage, simulation results are analyzed to evaluate the impact of different scenarios and make decisions based on the results.

Applications of Simulation in Decision Making

Simulation is used in a wide variety of fields for decision making. Some of the most common applications include:

Engineering Simulation: Simulation is used in engineering to model complex systems and predict their behavior in different situations. For example, it can be used to simulate fluid flow in a pipeline system, evaluate the performance of a production system, or predict the behavior of a structure during an earthquake.

Simulation in finance: Simulation is used in finance to model different economic scenarios and evaluate their impact on an investment portfolio or a company. For example, it can be used to evaluate investment risk, simulate the behavior of stock prices, or evaluate the impact of different investment strategies.

Simulation in healthcare: Simulation is used in healthcare to model complex biological systems and predict their behavior in different situations. For example, it can be used to simulate the behavior of a virus in a population, evaluate the impact of different treatments on patients, or simulate the behavior of an immune system.

Simulation in logistics: Simulation is used in logistics to model transportation systems and predict their behavior in different situations. For example, it can be used to simulate traffic behavior in a city, evaluate the impact of different transportation routes on product delivery, or simulate the behavior of a distribution system.

Simulation in energy: Simulation is used in energy to model energy systems and predict their behavior in different situations. For example, it can be used to simulate the behavior of an electric grid, evaluate the impact of different energy sources on the environment, or simulate the behavior of an energy storage system.

Advantages and disadvantages of simulation in decision-making

Simulation has several advantages and disadvantages that should be taken into account when using it in decision-making.

Advantages:

Allows modeling of complex systems: Simulation allows modeling of complex systems that would be difficult to understand or predict using analytical methods.

Allows simulation of different scenarios: Simulation allows simulation of different scenarios and evaluating the impact of each one. This allows making informed decisions on how to approach different situations.

Reduces risk: Simulation allows reducing risk by predicting the behavior of a system before making a decision. This can help avoid costly or dangerous errors.

Saves time and money: Simulation can save time and money by allowing different scenarios to be tested before implementing a solution.

Disadvantages:

Requires precise data: Simulation requires precise and complete data to build an accurate model. If the data is inaccurate or incomplete, the model will not accurately reflect the behavior of the real system.

Requires technical knowledge: Simulation requires technical knowledge to build models and run simulations. If the necessary experience is not available, errors can be made that affect the accuracy of the model.

Does not always reflect reality: Simulation is a simplification of the real world and there is always a degree of uncertainty associated with simulation models. Additionally, the assumptions used in the model may not be accurate and can affect the accuracy of the result.

Can be costly: Building an accurate simulation model can be costly, especially if specialized software or high-end hardware is required.

Overall, simulation can be a powerful tool for decision-making, especially when faced with complex or uncertain situations. However, it is important to recognize that simulation has limitations and that the results obtained must be interpreted carefully.

Stages of the Simulation Process

The simulation process consists of several stages that must be followed to obtain accurate and meaningful results. Below are the five main stages of the simulation process.

Problem Definition: In the first stage of the simulation process, the problem to be solved is defined and the simulation objectives are established. It is important to define the scope of the simulation and the limits of the model, including the assumptions used and the variables to be considered.

Model Construction: In the second stage of the simulation process, the mathematical model that describes the behavior of the system to be simulated is constructed. This involves defining the variables and relationships that influence the behavior of the system and developing the necessary equations to represent them.

Model Validation: In the third stage of the simulation process, the constructed model is validated to ensure that it is accurate and accurately represents the behavior of the real system. This involves comparing the simulation results with actual data to determine the accuracy of the model.

Simulation Execution: In the fourth stage of the simulation process, the simulation is executed using the input data defined in the first stage. Multiple simulations can be performed to evaluate different scenarios and obtain significant statistical results.

Results Analysis: In the fifth and final stage of the simulation process, the results obtained from the simulation are analyzed to evaluate the performance of the system in different situations and make informed decisions. The results should be compared with the objectives established in the first stage and the accuracy and reliability of the model should be evaluated.

Simulation Tools

There are several simulation tools available that can be used to construct and execute simulation models. Below are some of the most common tools used in simulation.

Discrete Event Simulators: Discrete event simulators are used to simulate systems that change state at discrete moments in time. This means that events occur at specific times and the system remains in a constant state between events.

Process Simulators: Process simulators are used to simulate systems that continuously change state over time. This implies that events occur continuously and the system does not remain in a constant state between events.

Monte Carlo Simulators: Monte Carlo simulators are used to simulate complex and stochastic systems using statistical methods. This technique uses random numbers to simulate variability and uncertainty in the model and provides probabilistic results.

System Dynamics Simulators: System dynamics simulators are used to simulate complex and dynamic systems that have multiple feedbacks and causal relationships. This technique uses flow diagrams to represent the system and can model the complexity of the interaction between different variables.

Agent-based Simulators: Agent-based simulators are used to simulate systems that consist of multiple individual agents that interact with each other. This technique is commonly used in modeling social and economic systems, such as simulating financial markets and decision-making in groups.

Each simulation tool has its own strengths and weaknesses, and the choice of tool depends on the type of problem being solved and the available data.

Examples of Simulation Applications:

Simulation is used in a wide variety of fields, from engineering and science to economics and social sciences. Below are some examples of simulation applications in different fields.

Engineering: Simulation is commonly used in engineering to design and optimize complex systems, such as airplanes, cars, and power plants. Simulation is used to evaluate the performance of different designs and to identify areas for improving efficiency and safety.

Health Sciences: Simulation is used in health sciences to evaluate the effect of different treatments and therapies on patients. Simulation is also used to train healthcare professionals in emergency situations and to develop and test new medical devices.

Economics and Finance: Simulation is used in economics and finance to model the behavior of financial markets and evaluate different investment strategies. Simulation is also used to evaluate the effect of different economic policies and to predict the impact of future events on the economy.

Social Sciences: Simulation is used in social sciences to model the behavior of individuals and groups in different situations. Simulation is used to evaluate the effect of different public policies and to predict the impact of future events on society.

Conclusion

Simulation is a powerful decision-making tool that is used in a wide variety of fields. Simulation can be used to evaluate different scenarios and make informed decisions in complex and uncertain situations. However, it is important to recognize that simulation has limitations and that the results obtained should be interpreted carefully. The simulation process consists of several stages that must be followed to obtain accurate and meaningful results, and there are several simulation tools available that can be used to build and execute simulation models.

SUPPLY CHAIN AND LOGISTICS MANAGEMENT

Supply chain and logistics management is one of the most important aspects of success for any business. It refers to the planning, coordination, and control of activities related to the procurement, manufacturing, and delivery of products or services to customers. In this chapter, we will explore the key concepts of supply chain and logistics management, including demand planning, inventory management, inbound and outbound logistics, and supply chain optimization.

Demand Planning:

Demand planning is the process of predicting the quantity of products or services that customers will purchase in the future. The accuracy of this prediction is crucial to ensure that a company has enough products in stock to meet customer demand without incurring additional inventory costs. To conduct effective demand planning, companies can use a variety of techniques, including collecting historical sales data, conducting customer opinion surveys, and analyzing market trends.

Inventory Management:

Inventory management is the process of managing a company's inventory to ensure that there are enough products in stock to meet customer demand without incurring unnecessary storage and management costs. Effective inventory management requires a combination of demand planning, forecasting, and supply chain management techniques. Companies can use a variety of tools and technologies to optimize their inventory management, including inventory management software, data analysis, and real-time inventory tracking systems.

Inbound Logistics:

Inbound logistics refers to the process of receiving and managing the materials and components needed for the production of products or services. Effective inbound logistics is essential to ensure that products are produced and delivered on time and at a reasonable cost. Companies can optimize their inbound logistics by implementing production planning systems, supply chain tracking, and collaboration with suppliers to improve efficiency.

Outbound Logistics:

Outbound logistics refers to the process of managing and delivering products or services to end customers. Effective outbound logistics is essential to ensure that products are delivered on time and in good condition, which can have a significant impact on customer satisfaction and brand loyalty. Companies can optimize their outbound logistics by implementing real-time shipping tracking systems, collaborating with logistics partners, and using information technology to improve supply chain visibility and efficiency.

Supply Chain Optimization:

Supply chain optimization is the process of identifying and eliminating inefficiencies and redundancies in a company's supply chain. Supply chain optimization can significantly improve efficiency, reduce costs, and improve customer satisfaction. Companies can optimize their supply chain through the implementation of information technology, such as the use of supply chain management systems and data analysis. They can also collaborate with suppliers and logistics partners to improve efficiency and transparency in the supply chain.

In addition, supply chain optimization may involve reorganizing processes and eliminating bottlenecks, which can improve efficiency throughout the supply chain. For example, a company could use lean techniques to identify and eliminate waste in the production and delivery of products.

Another common strategy for supply chain optimization is outsourcing non-core activities. For example, a company may outsource inbound logistics or inventory management to a specialized third party, allowing the company to focus on its core competencies and reduce costs.

Supply chain optimization may also involve improving collaboration and communication throughout the supply chain. Companies can work with suppliers and logistics partners to improve visibility and transparency in the supply chain, which can help prevent problems and delays in product delivery.

Conclusion:

In summary, supply chain and logistics management are key aspects of business success. Demand planning, inventory management, inbound and outbound logistics, and supply chain optimization are fundamental aspects of supply chain and logistics management. Companies can use a variety of tools and techniques to improve their supply chain management, including the use of information technology, collaboration with suppliers and logistics partners, and implementation of supply chain optimization strategies. By focusing on supply chain and logistics management, companies can improve efficiency, reduce costs, and improve customer satisfaction.

DESIGN OF FACILITIES AND PLANT LAYOUT

The design of facilities and plant layout are fundamental to the success of any company. The way production areas are organized, the location of equipment, and the efficiency of processes have a direct impact on the quality of products, delivery times, and operating costs.

This chapter covers the main concepts and strategies related to the design of facilities and plant layout, with a focus on resource optimization and continuous process improvement.

Design of facilities:

Facility design is the process of planning and configuring the physical spaces where productive activities take place. This includes the location of production areas, equipment, raw materials, and finished products, as well as the definition of workflow and space organization.

One of the main objectives of facility design is to optimize resource utilization, which involves maximizing efficiency and productivity of production processes, minimizing production times, and reducing operating costs. To achieve this, it is necessary to consider a number of factors such as the size and shape of spaces, the location of equipment and storage areas, and the layout of electrical, hydraulic, and ventilation facilities, among others.

Efficiency in facility design can significantly improve production processes. For example, proper planning of work areas can reduce material and product transportation times, resulting in greater efficiency and productivity. Likewise, the arrangement of production equipment can facilitate tasks and minimize the risk of work accidents.

Another important aspect of facility design is adaptability. Facilities must be flexible and able to be adapted to the changing needs of the company, depending on changes in demand, the introduction of new products, or the adoption of new technologies.

Plant layout:

Plant layout is the process of organizing and arranging production areas within an industrial plant. The purpose of plant layout is to achieve efficient interaction between available resources and production processes.

The main objective of plant layout is to reduce costs and improve production efficiency. To achieve this, it is necessary to consider a number of factors, such as the location of equipment, the layout of storage areas, the organization of workflow, and space planning.

Plant layout can also impact product quality and customer satisfaction. Proper arrangement of work areas can reduce production times, allowing for faster delivery of products with higher quality.

There are several approaches to plant layout design, including the process approach, which focuses on grouping activities by function, and the product approach, which groups production activities around the products being manufactured. There is also a hybrid approach, which combines both approaches to obtain the benefits of each.

The process approach is used when different products are produced using the same equipment and processes. In this case, production areas are grouped based on the activity being performed, such as machining, assembly, or finishing. This approach maximizes process efficiency, as the same equipment can be used to produce different products.

The product approach, on the other hand, is used when different products require specific processes and equipment. In this case, production areas are organized based on the products being manufactured. This approach optimizes production for each product, as specific processes and equipment can be tailored to each product.

The hybrid approach combines both approaches to obtain the benefits of each. In this case, production areas are grouped based on processes, but are adapted to the specific needs of each product.

Factors to consider in facility design and plant layout:

When designing facilities and laying out plants, it is necessary to consider a number of factors that have a direct impact on process efficiency and company profitability. Some of the most important factors to consider are:

Production capacity: It is necessary to determine the production capacity of the facilities to ensure that they are sufficient to meet current and future demand.

Workflow: It is necessary to analyze workflows to ensure that processes are performed efficiently and without interruptions.

Safety: It is necessary to ensure the safety of workers and the integrity of equipment, through the implementation of appropriate safety measures.

Ergonomics: It is necessary to consider ergonomics in the design of facilities and plant layout to avoid injuries or fatigue in workers.

Energy efficiency: It is necessary to consider energy efficiency in the design of facilities and plant distribution to reduce operating costs.

Flexibility: It is necessary to ensure that facilities are flexible and can be adapted to the changing needs of the company.

Strategies for continuous improvement in facility design and plant distribution:

Continuous improvement in facility design and plant distribution is crucial to maintaining efficiency and profitability in the company. Some strategies for continuous improvement in facility design and plant distribution include:

Implementation of quality management systems: Implementing quality management systems, such as ISO 9001, can help the company identify improvement opportunities in facility design and plant distribution.

Periodic evaluation of facilities: It is necessary to conduct periodic evaluations of facilities to identify improvement opportunities in plant distribution, workflow, safety, and energy efficiency.

Technology upgrades: Upgrading technologies can help improve process efficiency and reduce operating costs in facility design and plant distribution.

Worker training and education: It is necessary to ensure that workers are trained to use facilities efficiently and safely through continuous training and education.

Implementation of continuous improvement processes: Implementing continuous improvement processes, such as Lean Manufacturing or Six Sigma, can help identify and eliminate waste in production processes and improve efficiency in facility design and plant distribution.

Collaboration with suppliers and customers: Collaboration with suppliers and customers can help identify improvement opportunities in facility design and plant distribution and improve efficiency throughout the supply chain.

Conclusions:

Facility design and plant distribution are crucial to maintaining efficiency and profitability in the company. When designing facilities and distributing plants, it is necessary to consider a range of factors that have a direct impact on process efficiency and company profitability, such as production capacity, workflow, safety, ergonomics, energy efficiency, and flexibility.

Continuous improvement in facility design and plant distribution is crucial to maintaining efficiency and profitability in the company. Some strategies for continuous improvement in facility design and plant distribution include implementation of quality management systems, periodic evaluation of facilities, technology upgrades, worker training and education, implementation of continuous improvement processes, and collaboration with suppliers and customers.

In summary, facility design and plant distribution are critical tasks for the efficiency and profitability of the company. By considering the factors mentioned and implementing continuous improvement strategies, companies can significantly improve process efficiency and reduce operating costs, resulting in a competitive advantage in the market.

PRODUCTION PLANNING AND SCHEDULING

The chapter on Production Planning and Scheduling is one of the most important topics in production management, as it is responsible for establishing the necessary processes to achieve efficiency and effectiveness in the production of goods and services. This chapter covers the most relevant aspects of production planning and scheduling, its importance in business management, the methods and tools used, as well as the factors that influence its development and application.

Importance of Production Planning and Scheduling

Production planning and scheduling is essential for efficient production management. Its objective is to ensure that resources are available when needed to produce the required goods and services in the appropriate quantity and quality, at the lowest possible cost. This improves the profitability and competitiveness of the company.

Production planning and scheduling is a continuous process that begins with the definition of the company's objectives and goals, and extends to detailed production scheduling in the short term. Long-term objectives include defining the production strategy, identifying the markets and customers targeted by the company, defining the products and services offered, identifying the necessary resources, and planning production capacity.

In the short term, production planning and scheduling involves detailed scheduling of daily or weekly production, resource allocation to production tasks, production monitoring and control, and problem resolution and deviation management.

Methods and Tools for Production Planning and Scheduling

There are various methods and tools for production planning and scheduling, which are selected based on the characteristics and needs of the company and production. Some of the most commonly used methods and tools are:

Capacity Planning: This method involves evaluating available production capacity and identifying the resources needed to meet established production objectives. This allows the company to plan investment in resources necessary to increase production capacity and improve efficiency.

Production Scheduling: This tool allows for the allocation of resources to production tasks and the definition of start and end dates for activities. Production scheduling is carried out through computer tools such as production management systems (ERP) and advanced scheduling tools.

Production Control: Production control involves monitoring the progress of production and identifying deviations and problems. To do this, production tracking and control tools are used, such as Key Performance Indicators (KPIs), quality management systems, and Statistical Process Control (SPC) systems.

Factors Affecting Production Planning and Scheduling

Production planning and scheduling can be affected by various factors that influence its development and application. Some of these factors include:

Market demand: Market demand is a critical factor that affects production planning and scheduling. If market demand is high, the company must plan and schedule its production efficiently to ensure that it can meet market needs in the shortest possible time. On the other hand, if demand is low, the company must plan and schedule its production in a way that does not generate excess inventory that can result in unnecessary costs.

Resource availability: Resource availability, such as materials, machinery, and personnel, can influence production planning and scheduling. If resources are not available when needed, it can cause production delays and affect the company's ability to meet its delivery commitments.

Technical limitations: Technical limitations can influence production planning and scheduling. For example, if machinery cannot produce a certain amount of goods in a given time, the company must adjust its production plan accordingly.

Seasonality: Seasonality is an important factor that influences production planning and scheduling. For example, a company that produces toys may have very high demand during the holiday season and very low demand for the rest of the year. The company must plan and schedule its production to be able to meet demand during periods of high demand and avoid excess inventory during periods of low demand.

Competition: Competition is an important factor that influences production planning and scheduling. If competitors offer similar products at lower prices, the company must plan and schedule its production in a way that can compete on price and quality.

In summary, production planning and scheduling is a fundamental process in business management, as it allows for the optimization of resources and the improvement of efficiency and effectiveness in the production of goods and services. To achieve effective planning and scheduling, various factors that can affect the process must be considered, and appropriate tools and methods must be used to enable efficient and effective production management.

INVENTORY AND WAREHOUSE MANAGEMENT

Inventory and warehouse management is one of the most important activities in any company that sells products. Inventories represent a significant investment for the company and proper management can maximize profits and minimize losses. Warehouse management is the process of managing and organizing the products stored in a particular location. In this chapter, we will discuss the main strategies for inventory and warehouse management.

Inventory Management Strategies

Inventory management is the process of controlling the flow of products in a company. It is important to ensure that products are available to customers at the right time and in the right quantity. Below are the main strategies for inventory management.

Just-In-Time (JIT)

The JIT strategy involves having inventory just when it is needed. This means that the company does not maintain large quantities of inventory in stock, which reduces storage costs. The JIT strategy can be very effective for companies that produce high-demand products, as it minimizes inventory cost without sacrificing production capacity. However, this strategy is highly dependent on the supplier's ability to deliver products on time, which can be a risk.

Maximum-Minimum

The maximum-minimum strategy is based on establishing a maximum and minimum amount of inventory that must be maintained. When inventory reaches the minimum point, a replenishment order is placed. This strategy is useful for products that have a stable demand, as it allows for minimum inventory without risking running out of products. However, this strategy can lead to excess inventory if appropriate levels are not established.

ABC

The ABC strategy is based on classifying products into three categories: A, B, and C. A products are those with high demand and represent a large part of the company's sales. B products have moderate demand and C products have low demand. The company can apply different inventory management strategies for each category. For example, A products may have a higher inventory level than B and C products. This strategy allows the company to focus on the most important products and avoid excess inventory of low-demand products.

Warehouse Management Strategies

Warehouse management is the process of organizing the products stored in a particular location. Good warehouse management can improve efficiency and reduce storage costs. Below are the main strategies for warehouse management.

Fixed Location Storage

The fixed location storage strategy involves assigning a specific place for each product in the warehouse. This allows for more efficient inventory management and reduces the time needed to find a product. Additionally, this strategy can be very useful for companies that have a large number of different products.

Random Location Storage

The strategy of random location storage involves not assigning a specific location for each product in the warehouse. Instead, products are placed in any available space at the time they arrive at the warehouse. This strategy can be very effective for companies with a high product turnover rate, as it reduces the time required to find a specific product. However, this strategy can be more challenging to manage if the company has a large number of different products.

Cross-docking

The cross-docking strategy involves receiving products and directly shipping them to customers without storing them in the warehouse. This strategy can be very effective for companies with a high product turnover rate and an efficient distribution system. However, this strategy requires careful planning and efficient coordination between suppliers and customers.

Temperature Storage

The temperature storage strategy involves storing products at different temperatures according to their specific requirements. For example, perishable products may require cold storage, while electronic products may require dry and temperature-controlled storage. This strategy can be very effective for companies that sell a wide variety of products with different storage requirements.

Automated Storage

The automated storage strategy involves the use of automated systems to store and retrieve products in the warehouse. These systems can include robots, conveyors, and automated inventory control systems. This strategy can improve the efficiency and accuracy of the warehouse management process, but requires a significant investment in technology and training.

Conclusion

Inventory and warehouse management is a critical activity for any company engaged in the sale of products. Proper inventory management can maximize profits and minimize losses, while proper warehouse management can improve efficiency and reduce storage costs. The inventory and warehouse management strategies described in this chapter can help companies optimize their inventory and warehouse management according to their specific needs and requirements. It is important for companies to carefully evaluate these strategies and choose the ones that best fit their operations and business objectives.

METHODS AND TIME ENGINEERING

Methods and time engineering is a discipline that focuses on improving production processes through the identification and elimination of unnecessary activities and the optimization of necessary tasks to complete an activity or project. In this chapter, we will explain in detail what methods and time engineering involves, its objectives, its main techniques and tools, as well as its importance in the management of production processes and continuous improvement of organizations.

Definition of methods and time engineering

Methods and time engineering is defined as a discipline whose main objective is to improve production processes through the identification and elimination of unnecessary activities and the optimization of necessary tasks to complete an activity or project. This discipline focuses on efficiency management, cost reduction, and waste elimination in production processes.

Methods and time engineering is based on the systematic study of production processes to identify and eliminate all activities that do not add value, and optimize activities that do. Specific techniques and tools are used to measure, analyze and improve production processes.

Objectives of methods and time engineering

The objectives of methods and time engineering are as follows:

Efficiency improvement: Methods and time engineering aims to improve the efficiency of production processes by reducing the time required to complete an activity or project.

Cost reduction: Improving efficiency reduces the costs associated with completing an activity or project by eliminating unnecessary activities and optimizing necessary tasks.

Waste elimination: Methods and time engineering aims to eliminate waste in production processes, resulting in cost reduction and improvement in the quality of the product or service.

Quality improvement: Improving efficiency and eliminating waste leads to quality improvement in the product or service by eliminating errors and optimizing necessary tasks.

Techniques and tools of methods and time engineering

The main techniques and tools used in methods and time engineering are as follows:

Flowcharts: Flowcharts represent production processes graphically, identifying necessary activities and their relationship. These charts are useful for identifying unnecessary activities and optimizing necessary tasks.

Process analysis: Process analysis identifies and eliminates all activities that do not add value and optimizes necessary tasks. This analysis is conducted through direct observation of the production process and collection of relevant data, such as the time taken to complete each activity.

Time and motion study: The time and motion study is a technique that allows measuring the time required to perform a task or activity, identifying the necessary movements and eliminating those that do not add value. This technique is useful for improving efficiency and reducing costs associated with carrying out an activity or project.

Line balancing: Line balancing is a technique that allows optimally distributing the tasks necessary for the completion of a project or activity, in order to maximize efficiency and reduce costs. This technique is especially useful in chain production processes, where each task depends on the completion of the previous one.

Continuous improvement: Continuous improvement is a philosophy that focuses on the constant search for the improvement of production processes, through the identification and elimination of unnecessary activities and the optimization of necessary tasks. This philosophy is based on the belief that it is always possible to improve, and that continuous improvement is essential for the survival and growth of organizations.

Importance of methods and time engineering

Methods and time engineering is essential for the management of production processes and the continuous improvement of organizations, as it allows identifying and eliminating all activities that do not add value, and optimizing necessary tasks. Some of the reasons why methods and time engineering is important are:

Cost reduction: Improving efficiency and eliminating waste allows reducing costs associated with carrying out an activity or project, which translates into increased profitability for the organization.

Improvement in quality: Improving efficiency and eliminating errors allows improving the quality of the product or service, which translates into greater customer satisfaction and improvement in the reputation of the organization.

Greater flexibility: Optimizing production processes allows for greater flexibility in production management, which allows adapting to market needs and changes in demand.

Greater competitiveness: Improving efficiency and reducing costs allows for greater competitiveness in the market, which translates into a greater market share and an increase in the organization's profits.

Conclusions

In conclusion, methods and time engineering is an essential discipline for the management of production processes and the continuous improvement of organizations. This discipline allows identifying and eliminating all activities that do not add value, and optimizing necessary tasks for carrying out an activity or project. Techniques and tools used in methods and time engineering, such as flow charts, process analysis, time and motion study, line balancing, and continuous improvement, are essential for improving efficiency, reducing costs, improving quality, and competitiveness of organizations.

It is important to note that methods and time engineering is not a technique used only in the manufacturing industry but is also applicable in other areas, such as project management, logistics, customer service, among others. In any activity where a task needs to be performed, methods and time engineering can be applied to improve its efficiency and reduce associated costs.

In short, methods and time engineering is a discipline that provides great benefits to organizations, allowing for the improvement of production processes, cost reduction, and quality improvement, which translates into greater customer satisfaction and competitiveness in the market. Therefore, it is essential that organizations invest in the training and development of their teams in this discipline, to take advantage of its full potential and continue growing and improving in the future.

ERGONOMICS AND OCCUPATIONAL SAFETY

Ergonomics and occupational safety are two fundamental areas in the workplace. Ergonomics studies the relationship between the worker and their work environment with the aim of improving efficiency, productivity, and the worker's quality of life. On the other hand, occupational safety focuses on identifying, evaluating, and controlling risks that can affect the health and safety of workers.

This chapter will address the main aspects related to ergonomics and occupational safety. Basic concepts of ergonomics will be reviewed, as well as the main techniques and tools used for its application in the workplace. Additionally, the main regulations related to occupational safety will be analyzed, and the main strategies for preventing and controlling occupational risks will be discussed.

Ergonomics in the workplace

Ergonomics is a discipline that studies the relationship between the worker and their work environment with the aim of optimizing the worker's well-being, increasing efficiency and productivity, and reducing the risk of occupational injuries and illnesses.

The first step to applying ergonomics in the workplace is to conduct a detailed analysis of the tasks and activities carried out in the company. This includes observing workers in their work environment, as well as evaluating the equipment, tools, and furniture used.

Once the analysis has been conducted, the following ergonomics techniques and tools can be applied in the workplace:

Ergonomic design of the workstation: The ergonomic design of the workstation involves adapting the work environment to the worker with the aim of minimizing the risk of occupational injuries and illnesses. This includes selecting ergonomic furniture and equipment, optimizing lighting and temperature, and adapting the workspace to meet the needs of the worker.

Workload assessment: Workload assessment involves identifying tasks and activities that require greater physical or mental effort with the aim of designing a work environment that reduces fatigue and stress in the worker.

Biomechanical analysis: Biomechanical analysis involves evaluating how the human body moves and exerts itself in different work situations with the aim of designing a work environment that minimizes the risk of musculoskeletal injuries.

Posture analysis: Posture analysis involves evaluating how the worker sits, stands, or moves in the work environment with the aim of designing a work environment that promotes a healthy posture and reduces the risk of musculoskeletal injuries.

Ergonomics training: Ergonomics training involves educating workers about the basic principles of ergonomics with the aim of promoting a healthy work culture and preventing occupational injuries. Workers can learn about the importance of proper posture, how to adjust the height of the desk and chair, how to use ergonomic equipment, and how to take breaks to stretch and rest.

Safety regulations and standards in the workplace

Workplace safety is a discipline that focuses on identifying, evaluating, and controlling risks that may affect the health and safety of workers. To this end, there are various regulations and standards that establish the obligations and responsibilities of employers and workers in the field of workplace safety.

Some of the most important workplace safety regulations and standards are:

Regulations on Health and Safety at Work: These regulations establish the obligations and responsibilities of employers and workers in the field of workplace safety. It also sets criteria for the identification, evaluation, and control of occupational risks.

Regulations on Safety and Health at Work: These regulations establish provisions for the prevention of occupational accidents and illnesses. They also set out procedures and criteria for evaluating occupational risks and implementing preventive measures.

Strategies for preventing and controlling occupational risks

Preventing and controlling occupational risks is essential to ensure a healthy and safe work environment. There are various strategies that can be implemented by employers and workers, including:

Risk assessment and management: Risk assessment and management involves identifying occupational risks, evaluating their likelihood of occurrence and severity, and implementing preventive measures to minimize or eliminate these risks.

Training and education: Training and education of workers on workplace safety are essential to promote a culture of safe work. Workers should be trained in the use of personal protective equipment, identifying occupational risks, and implementing preventive measures.

Implementation of safety measures: Safety measures are technical or administrative measures that are implemented to reduce or eliminate occupational risks. These measures may include the installation of collective protection equipment, the implementation of safe work procedures, and the use of personal protective equipment.

Safety inspections: Safety inspections are an important tool for identifying occupational risks and evaluating the effectiveness of preventive measures implemented. Inspections can be carried out by employers or workers and should be recorded and reported for follow-up.

Conclusions

Ergonomics and occupational safety are fundamental aspects for ensuring a healthy and safe work environment. Ergonomics focuses on the design and adaptation of the work environment to the physical and psychological needs of workers, while occupational safety focuses on identifying, evaluating, and controlling risks that may affect the health and safety of workers.

It is important for employers and workers to be aware of the importance of ergonomics and occupational safety and work together to implement preventive measures and ensure a safe work environment. This involves identifying and evaluating occupational risks, implementing preventive measures, providing training and education to workers, and conducting safety inspections.

Additionally, it is important for employers to comply with the safety and health regulations established by competent authorities. The General Law on Health and Safety at Work, the NOMs, and the Federal Regulation on Safety and Health at Work establish the obligations and responsibilities of employers and workers in terms of occupational safety and establish criteria for identifying, evaluating, and controlling occupational risks.

In conclusion, ergonomics and occupational safety are fundamental aspects for ensuring a healthy and safe work environment. Employers and workers must work together to implement preventive measures and ensure a safe work environment and comply with the safety and health regulations established by competent authorities.

MAINTENANCE MANAGEMENT AND RELIABILITY

Maintenance management and reliability are critical aspects for the success of any organization. The effective implementation of these processes can improve operational efficiency, reduce maintenance costs, increase asset lifespan, and enhance personnel and machinery safety. In this chapter, we will explore the basics of maintenance management and reliability, their benefits, and how they can be implemented in an organization.

Maintenance Management

Maintenance management refers to the processes used to maintain and improve the functionality, reliability, and safety of equipment and systems in an organization. Maintenance management can be divided into two main categories: corrective maintenance and preventive maintenance.

Corrective maintenance is performed after a piece of equipment or system has failed. The main objective of corrective maintenance is to repair the equipment or system as quickly as possible to minimize downtime. However, corrective maintenance can be expensive and may have a negative impact on an organization's productivity and profitability.

Preventive maintenance, on the other hand, is performed before a failure occurs in equipment or system. The main objective of preventive maintenance is to prevent failures in equipment or system and minimize downtime. Preventive maintenance can be divided into two categories: time-based maintenance and condition-based maintenance.

Time-based maintenance is performed at regular intervals, regardless of the equipment or system's condition. This type of maintenance is useful for equipment and systems that have a predictable lifecycle, such as changing a car's oil after a certain number of kilometers.

Condition-based maintenance, on the other hand, is performed when the equipment or system reaches a certain level of degradation or when an anomaly is detected. This type of maintenance is based on continuous monitoring of equipment and systems to identify issues before failures occur.

Reliability

Reliability refers to the ability of equipment or system to operate continuously and without failures for a certain period of time. Reliability is a critical aspect of any equipment or system, especially in high-security and mission-critical environments. Reliability can be improved by implementing maintenance management processes and applying reliability analysis techniques.

Failure rate is a common measure of reliability. Failure rate refers to the frequency with which failures occur in equipment or system during a certain period of time. Failure rate can be reduced by implementing preventive maintenance processes, using high-quality materials, and improving equipment or system design.

The Mean Time Between Failures (MTBF) is another common measure of reliability in industrial engineering. The MTBF refers to the average time between two consecutive failures in a piece of equipment or system. The MTBF can be increased by improving the equipment or system design, using high-quality materials, and implementing preventive maintenance processes.

Another important aspect of reliability is availability. Availability refers to the time during which a piece of equipment or system is available and operational. Availability can be improved by implementing preventive maintenance processes, reducing repair time, and improving the equipment or system design.

Implementation of maintenance and reliability management

The effective implementation of maintenance and reliability management can significantly improve operational efficiency, reduce maintenance costs, and enhance personnel and machinery safety. Here are some key steps for successfully implementing maintenance and reliability management in an organization:

Evaluate the current situation: The first step in implementing maintenance and reliability management is to evaluate the current situation of the organization. This may include reviewing existing maintenance records, identifying current problems, and assessing equipment or system performance.

Develop a maintenance plan: Once the current situation has been evaluated, it is important to develop a maintenance plan that addresses identified problem areas. The maintenance plan should include a combination of preventive and corrective maintenance to minimize downtime and improve equipment or system reliability.

Implement preventive maintenance processes: Preventive maintenance is fundamental to improving equipment or system reliability. Preventive maintenance processes should be implemented to ensure that equipment or systems are kept in optimal condition.

Use reliability analysis techniques: Reliability analysis techniques such as Failure Mode and Effect Analysis (FMEA) and Fault Tree Analysis (FTA) can help identify areas of highest risk and develop mitigation strategies.

Train personnel: Personnel should be trained in maintenance processes and reliability analysis techniques. This will ensure that personnel are prepared to perform preventive maintenance and identify potential problems before failures occur.

Monitor performance: Once maintenance and reliability management processes are implemented, it is important to monitor equipment or system performance. This may include monitoring failure rates, MTBF, and availability.

Benefits of Maintenance and Reliability Management

The effective implementation of maintenance and reliability management can provide several significant benefits for an organization. Some of these benefits include:

Reduction of maintenance costs: The implementation of preventive maintenance processes can reduce long-term maintenance costs by minimizing the need for costly and extensive repairs.

Improvement of operational efficiency: The implementation of preventive maintenance processes can improve operational efficiency by minimizing downtime and increasing equipment or system availability.

Increase in asset lifespan: The implementation of preventive maintenance processes can increase asset lifespan by minimizing equipment or system wear and fatigue.

Improvement of safety: The implementation of preventive maintenance processes can improve personnel and machinery safety by identifying and correcting potential issues before failures occur.

Improvement of product quality: The implementation of preventive maintenance processes can improve product quality by reducing variability in the production process and minimizing product defects.

Improvement of customer satisfaction: Improving operational efficiency and product quality can improve customer satisfaction by providing high-quality products in a faster delivery time.

Conclusions

In summary, maintenance and reliability management are fundamental to operational efficiency, safety, and product quality. The effective implementation of maintenance and reliability management can reduce maintenance costs, improve operational efficiency, increase asset lifespan, improve safety, improve product quality, and improve customer satisfaction. Preventive maintenance processes, reliability analysis techniques, and personnel training are some of the key aspects of maintenance and reliability management. By implementing these processes, organizations can achieve greater reliability and operational efficiency in their production.

AUTOMATION AND CONTROL TECHNOLOGIES

Currently, automation and control technology is present in a wide variety of industries, from manufacturing production to the management of infrastructure and public services. This technology uses control systems to automate processes and improve the efficiency, quality, and safety of the products and services offered.

In this chapter, we will explore the main automation and control technologies, as well as the benefits and challenges they present. In addition, we will examine the different types of control systems, from relay-based control systems to more advanced programmable control systems.

Automation and Control Technologies

Automation and control technologies include a wide range of tools and systems designed to improve the efficiency, quality, and safety of industrial processes. These technologies are used in a variety of industries, from manufacturing to the management of infrastructure and public services.

Some of the main automation and control technologies include:

Computer Numerical Control (CNC)

Computer Numerical Control (CNC) is an automation system used in manufacturing production to control machine tools using software programs. CNC software programs allow users to define the tools and movements that should be performed on the workpiece.

CNC is widely used in the production of high-precision parts, such as components for the aerospace and medical industries.

Robotics

Robotics is an automation technology that uses robots to perform tasks in a variety of industrial environments. Robots can be programmed to perform repetitive and dangerous tasks, which increases worker safety and improves process efficiency.

Robotics is used in a wide range of industries, from manufacturing to logistics and healthcare.

Process Control Systems

Process control systems are automation tools used to control and monitor industrial processes in real time. These systems use sensors and controllers to measure and adjust process parameters, such as temperature, pressure, and flow.

Process control systems are used in a wide range of industries, from chemical production to food and beverage manufacturing.

Quality Control Systems

Quality control systems are automation tools used to ensure that products and services meet required quality standards. These systems use measurement and analysis techniques to detect and correct quality problems.

Quality control systems are used in a wide range of industries, from manufacturing to healthcare and financial services.

Types of Control Systems

There are several types of control systems used in automation and control technology. These systems are classified based on their complexity and programming capability.

Relay-based Control Systems

Relay-based control systems are the oldest and simplest control systems. These systems use electromechanical relays to control the operation of a machine or process. Relays are activated or deactivated based on the electrical signal they receive, allowing for control of the on/off operation of system components.

Relay-based control systems are limited in their programming capabilities and can only perform simple and repetitive tasks. However, these systems are reliable and are used in applications where simplicity is more important than advanced functionality.

Programmable Logic Controller (PLC)

Systems Programmable logic controller (PLC) systems are a more advanced form of control that uses a programmable computer to control the operation of a machine or process. PLCs use specialized programming language to control the sequence of operations and adjust process parameters.

PLCs are flexible and can be programmed to perform a wide variety of tasks and operations. These systems are used in a wide range of industrial applications, from manufacturing production to management of infrastructure and public services.

Distributed Control Systems (DCS)

Distributed control systems (DCS) are advanced control systems used in complex industrial processes. These systems use a network of distributed controllers to monitor and control multiple processes in real-time.

DCSs are highly scalable and can be programmed to control large-scale and complex processes. These systems are used in a wide range of industrial applications, from chemical production to power generation and infrastructure management.

High-Level Control Systems (HIL)

High-level control systems (HIL) are simulation systems used to test and validate control systems in virtual environments. These systems use mathematical models to simulate system behavior in real-time, allowing engineers to test and optimize control system performance before implementing them in a real-world environment.

HIL systems are used in a wide range of industrial applications, from automotive to aerospace and electrical engineering.

Benefits of Automation and Control Technology

Automation and control technology offer several benefits to companies and organizations that use it. Some of the main benefits include:

Increased Efficiency

Automation and control allow companies to improve the efficiency of their processes, reducing production time and minimizing human errors. This allows companies to increase their production and reduce manufacturing costs.

Improved Quality

Automation and control technology allows companies to improve the quality of their products and services, reducing variability and minimizing errors. This increases customer satisfaction and improves the brand image.

Increased Safety

Automation and control technology allows for improved workplace safety by minimizing human exposure to hazardous situations. This is achieved by automating hazardous processes or by implementing safety systems that monitor the operation of machines and equipment to detect faults or dangerous conditions.

Cost Reduction

Automation and control technology can help reduce production costs by minimizing waste of materials and energy, as well as reducing the need for human labor.

Increased Flexibility

Advanced control systems, such as PLCs and DCSs, are highly flexible and can be programmed to perform a wide variety of tasks and operations. This allows companies to quickly adapt to changes in the market and new customer demands.

Improved Data Analysis

Automation and control technology allows companies to collect and analyze large amounts of data about their processes and operations. This enables them to identify areas for improvement and optimize their processes to maximize efficiency and profitability.

Challenges in Implementing Automation and Control Technology

While automation and control technology offers a number of benefits to companies, it also presents a number of challenges in its implementation. Some of the main challenges include:

Cost

The implementation of advanced control systems can be costly, especially for small and medium-sized enterprises. In addition to the cost of equipment and control systems, highly trained personnel are also required to install, program, and maintain these systems.

Integration

The implementation of advanced control systems often requires the integration of multiple systems and equipment, which can be a technical challenge. In addition, control systems may not be compatible with existing systems, requiring additional investment in updating existing systems or acquiring new equipment.

Personnel Training

The implementation of advanced control systems requires highly trained and specialized personnel to program, maintain, and operate the systems. Personnel training can be costly and time-consuming, which can delay the implementation of the control system.

Cybersecurity

Advanced control systems are connected to the network and can be vulnerable to cyber attacks. Cybersecurity must be an important consideration in the implementation of advanced control systems to protect systems and sensitive data.

Examples of Automation and Control Technology in Industry

Automation and control technology is used in a wide range of industries, from manufacturing to energy and automotive. The following are some examples of how automation and control technology is used in industry.

Automotive

In the automotive industry, automation and control technology is used to improve efficiency and quality in car production. Advanced control systems are used to control the production process, from part assembly to painting and final finishing. Robot and automated machine systems are also used to perform tasks that previously required human labor, such as component assembly and welding.

Energy

In the energy industry, automation and control technology is used to improve efficiency and safety in energy production, from electricity generation to oil and gas extraction. Advanced control systems are used to monitor and control the operation of power plants, which helps optimize production and reduce costs.

Manufacturing

In the manufacturing industry, automation and control technology is used to improve efficiency and quality in the production of goods, from electronics manufacturing to food and beverage production. Advanced control systems are used to monitor and control production processes, which helps to reduce costs and improve product quality.

Mining

In the mining industry, automation and control technology is used to improve safety and efficiency in the extraction of minerals and metals. Advanced control systems are used to monitor and control the operation of mining equipment, which helps to minimize risks for workers and maximize production.

Agriculture

In the agricultural industry, automation and control technology is used to improve efficiency and productivity in the production of food and crops. Advanced control systems are used to monitor and control processes such as irrigation, fertilization, and harvesting, which helps to maximize crop yields and reduce production costs.

Conclusions

Automation and control technology is a powerful tool for improving efficiency, safety, and profitability in a wide range of industries. The implementation of advanced control systems can help companies to reduce costs, improve product quality, and quickly adapt to changes in the market.

However, the implementation of advanced control systems also presents a number of challenges, such as cost, integration, personnel training, and cyber security. It is important for companies to carefully evaluate the benefits and challenges of implementing advanced control systems before making a decision.

Ultimately, automation and control technology can be a valuable tool to help companies remain competitive in an increasingly demanding global market.

INDUSTRIAL PROJECT DEVELOPMENT AND MANAGEMENT

The development and management of industrial projects is a crucial task for companies seeking to improve their production processes and, therefore, increase their competitiveness in the market. This chapter aims to provide an overview of the processes and tools necessary to successfully carry out industrial projects.

Phases of industrial project development

The development of an industrial project consists of several phases, which may vary depending on the nature of the project and the company carrying it out. However, it is possible to identify some phases that are common to most projects:

Identification of the need: In this phase, the problem or need to be solved through the project is defined. The project's objectives are also identified, and an estimated timeframe and budget are established.

Planning: In this phase, a detailed plan is developed that establishes the activities necessary to achieve the project's objectives, the resources that will be needed, and the deadlines in which they will be carried out.

Execution: In this phase, the planned activities are carried out, and the results obtained are monitored and controlled.

Closure: In this phase, the results obtained are evaluated, the project is documented, and the necessary actions are taken to close it.

Tools for industrial project management

There are various tools that can be used to carry out the management of industrial projects. Some of the most common ones are:

Gantt chart: This tool allows the graphical representation of the project's activity plan, showing the tasks to be performed and their estimated duration. It also allows the establishment of dependency relationships between tasks.

Pert Network: This tool is similar to the Gantt chart, but focuses on the dependency relationships between tasks. It allows identifying which are the critical tasks of the project, i.e., those that must be completed on time for the project to be completed on time.

Risk matrix: This tool allows identifying the risks associated with the project and establishing strategies to minimize or eliminate them. It also allows evaluating the probability and impact of each risk.

Contingency plan: This plan establishes the measures that will be taken in case of any unforeseen event that affects the project's development. It is important that this plan is established before the unforeseen event occurs, to be able to act quickly and effectively.

Project management software: There are various computer tools that allow the efficient management of projects, such as Microsoft Project or Trello. These tools allow planning tasks, establishing deadlines and resources, and monitoring the project's development in real-time.

Key factors for the success of an industrial project

The success of an industrial project depends on various factors, some of which are:

Clear definition of the project's objectives: It is important that the project's objectives are well defined from the beginning, so that all team members work in the same direction and know what the expected outcome is.

Proper resource allocation: it is crucial to assign the appropriate resources to the project, including personnel, budget, and time. Failure to allocate necessary resources may cause difficulties and delays in the project.

Effective communication: it is essential to have clear and effective communication among all members of the project team, as well as with clients and suppliers. Good communication can prevent misunderstandings and problems that could affect the project's success.

Risk management: it is important to identify the risks associated with the project and establish strategies to minimize or eliminate them. This can help prevent problems and delays in the project's development.

Monitoring and evaluation: it is necessary to constantly monitor the project to ensure that it is progressing according to the established plan and to detect any deviations or issues. Additionally, it is important to conduct an evaluation at the end of the project to determine if objectives have been met and expected results have been obtained.

Example of Industrial Project Management

To illustrate the management of an industrial project, the following hypothetical example is presented:

Company X wants to implement a new production line to manufacture a specific product. The project will be completed within six months, and a budget of $100,000 has been established. The project team consists of a project manager, a production engineer, an industrial designer, and a team of production operators.

Phase 1: Needs identification

The project team meets to define the problem they want to solve and the project's objectives. It is established that the company needs a new production line to manufacture a specific product due to increased demand. The project's objectives are to increase production of the product, improve quality, and reduce costs.

Phase 2: Planning

The project team develops a detailed plan that establishes the activities necessary to achieve the project's objectives, the resources required, and the deadlines for completion. The following main activities are established:

Production line design

Acquisition of necessary equipment

Production operator training

Production testing and adjustments

The project team uses a Gantt chart to graphically represent the project's activity plan and establish task dependencies.

Phase 3: Execution

The project team begins to carry out the planned activities and tracks and controls the results obtained. Regular meetings are established to assess project progress and make adjustments if necessary.

Phase 4: Closure

Once the implementation of the new production line is completed, the project is evaluated to determine if objectives have been met and expected results have been obtained. The project is documented, and closure is completed.

Conclusion

Managing industrial projects is a complex task that requires detailed planning, proper resource management, effective communication, risk identification and management, and constant project monitoring. The success of the project will depend largely on proper management of these key areas.

It is important for industrial project managers to have a good understanding of the activities and processes that will be carried out in the project, as well as the resources required to complete it. They also need leadership, communication, and time management skills to lead the project team and ensure that deadlines and objectives are met.

The use of project management tools, such as Gantt charts, can be very helpful in planning and monitoring industrial projects. These tools provide a clear visualization of activities and timelines and can help identify bottlenecks and delays in the project.

In summary, managing industrial projects is a complex process that requires careful planning, proper resource and risk management, effective communication, and constant project monitoring. Proper implementation of these practices can help ensure project success and client satisfaction.

COST-BENEFIT ANALYSIS AND PROJECT EVALUATION

In project management, it is essential to conduct a careful evaluation of costs and benefits before making any significant decisions. This is because any investment in a project must be profitable and beneficial to the organization as a whole. To evaluate the costs and benefits of a project, a technique called cost-benefit analysis is used. In this chapter, we will discuss in detail what cost-benefit analysis is, how it is carried out, and its importance in project evaluation.

What is Cost-Benefit Analysis?

Cost-benefit analysis is a technique used to evaluate the relationship between the costs and benefits of a project. This analysis is carried out to determine whether the expected benefits of the project justify the costs associated with it. In other words, cost-benefit analysis helps organizations determine whether a project is profitable and economically viable.

Cost-benefit analysis involves the identification of all costs associated with the project, including both direct and indirect costs. Direct costs are those that are directly related to the execution of the project, such as material and labor costs. Indirect costs, on the other hand, are those that are related to the project but are not directly attributable to it, such as general administrative costs.

On the other hand, project benefits must be identified and quantified. Benefits can be tangible or intangible. Tangible benefits are those that can be measured in monetary terms, such as increased sales or cost reduction. Intangible benefits are those that cannot be easily measured in monetary terms, such as brand image improvement or customer satisfaction.

Once all costs and benefits associated with the project have been identified, an analysis is performed to determine whether the benefits justify the costs. If the analysis shows that the benefits outweigh the costs, then the project is considered economically viable.

Steps to Conduct Cost-Benefit Analysis

Cost-benefit analysis is carried out in several stages. These stages include:

Identification of Costs and Benefits: The first step in cost-benefit analysis is to identify all costs and benefits associated with the project. This includes both direct and indirect costs and tangible and intangible benefits.

Quantification of Costs and Benefits: The next step is to quantify all identified costs and benefits. Costs are quantified in monetary terms, while tangible benefits are also quantified in monetary terms. Intangible benefits are quantified using subjective methods such as surveys or opinion analysis.

Establishment of a Baseline: Once all costs and benefits have been identified and quantified, a baseline is established. The baseline is a representation of the expected costs and benefits without the project in question. This helps to compare the expected costs and benefits with the actual costs and benefits after project implementation.

Cost-Benefit Analysis: The next step is to carry out a real cost-benefit analysis. This involves comparing the expected costs and benefits with the established baseline. If the expected benefits outweigh the expected costs, then the project is considered economically viable. If the expected costs outweigh the expected benefits, then the project may need to be re-evaluated or abandoned.

Importance of Cost-Benefit Analysis in Project Evaluation

Cost-benefit analysis is an important tool for project evaluation for several reasons:

Helps make informed decisions: Cost-benefit analysis provides important information about the relationship between the costs and benefits of a project.

This helps decision-makers to make informed decisions about whether to proceed with a project or not.

Identifies hidden costs: Cost-benefit analysis helps to identify hidden costs associated with a project. This includes indirect costs that are often overlooked but can have a significant impact on the profitability of the project.

Improves project planning: Cost-benefit analysis helps organizations to plan and budget a project effectively. This helps to avoid unpleasant surprises and ensure that project objectives are achieved within the allocated budget.

Reduces risk: Cost-benefit analysis helps organizations to evaluate the risk associated with a project. This includes the assessment of financial, operational, and compliance risks. If a high risk is identified, then steps can be taken to reduce or mitigate that risk before implementing the project.

Example of Cost-Benefit Analysis

To illustrate how a cost-benefit analysis is performed, let's consider a hypothetical example of an organization that is considering implementing a new inventory management system. Suppose the organization has identified the following costs and benefits associated with the project:

Costs:

Software cost: $20,000

Implementation cost: $5,000

Training personnel cost: $2,000

Additional operating costs: $1,000 per month

Benefits:

Inventory cost reduction: $3,000 per month

Increased sales: $2,000 per month

Inventory error reduction: $1,000 per month

Improved customer satisfaction: not quantifiable

Using this information, we can carry out a cost-benefit analysis in the following way:

Identification of costs and benefits: The identified costs include the cost of the software, the cost of implementation, the cost of training personnel, and additional operating costs. The identified benefits include a reduction in inventory costs, an increase in sales, and a reduction in inventory errors. An improvement in customer satisfaction is also identified, although it cannot be quantified.

Establishment of a baseline: The baseline is established by identifying the costs and benefits that are expected without the implementation of the project. Suppose that, without the new inventory management system, the organization expects to have inventory costs of $10,000 per month, sales of $20,000 per month, and inventory errors of $2,000 per month.

Identification of actual costs and benefits: Once the new inventory management system is implemented, the actual reduction in inventory costs, actual increase in sales, and actual reduction in inventory errors can be measured. Suppose these benefits turn out to be $4,000 per month, $2,500 per month, and $1,500 per month, respectively. Additionally, the additional operating costs turn out to be $1,500 per month.

Cost-benefit analysis: Using this information, we can compare the expected costs and benefits with the baseline and the actual costs and benefits to determine if the project is economically viable. The cost-benefit analysis is done in the following way:

Expected costs: $20,000 + $5,000 + $2,000 + ($1,000 x 12) = $39,000

Expected benefits: $3,000 + $2,000 + $1,000 = $6,000

Actual costs: $39,000 + ($1,500 x 12) = $57,000

Actual benefits: $4,000 + $2,500 + $1,500 = $8,000

Based on this analysis, we can see that the expected benefits exceed the expected costs, indicating that the project is economically viable. Furthermore, the actual benefits also exceed the actual costs, suggesting that the implementation of the new inventory management system has been a success.

Limitations of cost-benefit analysis

Although cost-benefit analysis is a useful tool for project evaluation, there are some important limitations that must be taken into account:

Difficulty in quantifying certain benefits: It can be difficult to quantify certain benefits, such as improvement in customer satisfaction. This can make it more difficult to determine if a project is economically viable.

Inaccurate assumptions: Cost-benefit analysis is based on assumptions, and if these assumptions are inaccurate, then the analysis can be incorrect. Therefore, it is important to ensure that the assumptions used are accurate and realistic.

Does not account for long-term costs and benefits: Cost-benefit analysis focuses on short-term costs and benefits and does not account for long-term costs and benefits. Therefore, a project may appear economically viable in the short term but not in the long term.

Conclusion

In summary, cost-benefit analysis is an important tool for project evaluation. It allows managers and decision-makers to determine whether a project is economically viable by comparing expected costs and benefits with actual costs and benefits. However, it is important to consider the limitations of cost-benefit analysis, such as the difficulty in quantifying certain benefits and inaccurate assumptions.

In addition to cost-benefit analysis, there are other tools and techniques that can be used for project evaluation, such as cost-effectiveness analysis, cost-utility analysis, and environmental and social impact analysis. Each of these tools and techniques has its own advantages and limitations, and the choice of the appropriate tool will depend on the project objectives and specific circumstances.

Ultimately, project evaluation is essential to ensure that organizations invest their resources effectively and efficiently. By evaluating the costs and benefits of a project, managers and decision-makers can make informed decisions about whether to continue with the project or not. This can help ensure that limited resources are used effectively to achieve the organization's goals and maximize its impact.

FINANCIAL ANALYSIS AND PROFITABILITY IN INDUSTRIAL ENGINEERING

Financial analysis and profitability analysis are fundamental tools in industrial engineering, as they allow for the evaluation of project viability and profitability, as well as strategic and tactical decision-making within a company. In this chapter, we will discuss the key concepts and techniques used in financial analysis and profitability, as well as their importance in industrial engineering.

Fundamental concepts

Before delving into financial analysis and profitability analysis, it is important to understand some fundamental concepts, such as the cost of capital, cash flow, discount rate, and break-even point.

The cost of capital refers to the rate of return required by investors to finance a project. This rate may be comprised of different components, such as the cost of debt, the cost of equity, and the cost of assets.

Cash flow refers to the money that enters and leaves a company during a given period. It is important to distinguish cash flow from accounting profits, as accounting profits do not always translate to cash flow.

The discount rate is the interest rate used to calculate the present value of future cash flows. This rate is used to evaluate the profitability of a project.

The break-even point refers to the sales level at which revenue equals costs. This point is important because it indicates the minimum sales level needed to avoid losses.

Financial analysis

Financial analysis refers to the study of a company's financial statements, such as the balance sheet, income statement, and cash flow statement. These financial statements provide important information about a company's financial position and performance.

The balance sheet shows the assets, liabilities, and equity of a company at a specific point in time. Assets are the resources that the company possesses, such as cash, accounts receivable, inventory, and property. Liabilities are the company's obligations, such as accounts payable and loans. Equity is the shareholders' investment in the company.

The income statement shows a company's revenues and expenses during a specific period. Revenues include sales of products or services, while expenses include production costs, administrative expenses, and taxes.

The cash flow statement shows a company's cash flows during a specific period. This financial statement is important because it shows the actual cash flow of the company, which allows for evaluating its ability to finance its operations and investments.

Different financial ratios, such as liquidity, profitability, and leverage, are used to analyze a company's financial statements.

Liquidity refers to a company's ability to pay its short-term debts. Different liquidity ratios are used, such as the current ratio and the quick ratio.

Profitability refers to a company's ability to generate profits from its operations. Different profitability ratios are used, such as return on equity, return on assets, and profit margin.

Leverage refers to a company's level of debt in relation to its assets and equity. Different leverage ratios are used, such as the debt ratio and the interest coverage ratio.

It is important to note that financial ratios should not be evaluated in isolation but should be analyzed in conjunction with the company's goals and financial situation.

Profitability analysis

Profitability analysis refers to evaluating the profitability of a project or investment. Different techniques are used in this type of analysis, such as net present value (NPV), internal rate of return (IRR), and payback period.

Net present value is a technique used to evaluate the profitability of an investment over time. NPV is calculated by subtracting the investment cost from the future cash flows discounted at an appropriate discount rate. If NPV is positive, the investment is profitable.

Internal rate of return is another technique used to evaluate the profitability of an investment. IRR is the discount rate at which NPV is equal to zero. If IRR is greater than the required rate of return, the investment is profitable.

Payback period refers to the time required to recover the initial investment. It is calculated by dividing the investment cost by the expected annual cash flow. If the payback period is less than the projected project time, the investment is profitable.

It is important to note that profitability analysis should consider not only expected cash flows but also investment risks, such as uncertainty in revenues and costs and changes in market conditions.

Application in Industrial Engineering

Financial and profitability analysis is a fundamental tool in industrial engineering, as it allows for the evaluation of project feasibility and profitability, as well as strategic and tactical decision making within a company.

In industrial engineering, financial and profitability analysis is applied in different areas, such as investment project evaluation, cost management, and financial decision making.

In investment project evaluation, profitability analysis is used to assess the feasibility of projects and determine if they are profitable or not. Additionally, financial analysis is used to identify the costs and revenues associated with projects and evaluate the sensitivity of results to different assumptions.

In cost management, financial analysis is used to identify the fixed and variable costs of a company and evaluate the profitability of its products and services. Different techniques, such as cost-volume-profit analysis, are also used to evaluate the impact of changes in costs and prices on the company's profitability.

In financial decision making, financial and profitability analysis is used to evaluate different options and determine the most profitable one. For example, different financing options, such as bond issuance or bank loans, can be evaluated to determine which option is the most profitable for the company in terms of costs and risks.

Furthermore, financial and profitability analysis is also used in evaluating the efficiency and effectiveness of a company's operations. For instance, different production options, such as outsourcing certain operations or automating processes, can be evaluated to determine the most profitable option in terms of costs and productivity.

In summary, financial and profitability analysis is a fundamental tool in industrial engineering, as it allows for the evaluation of project feasibility and profitability, as well as strategic and tactical decision making within a company. The application of financial and profitability analysis techniques enables companies to make informed decisions and minimize the risks associated with their operations and projects.

LEGAL ASPECTS AND REGULATIONS IN THE INDUSTRY

Industry is one of the main economic drivers of any country. However, the production of goods and services cannot be carried out without complying with a series of regulations and regulations established by government authorities and corresponding regulatory agencies. In this chapter, legal aspects and regulations applicable to the industry will be analyzed, in order to better understand the legal and regulatory framework that governs this sector.

Legal framework

The legal framework that regulates the industry varies from country to country and according to different legislations. However, there are certain general aspects that are common in most countries. One of the most important legal aspects is intellectual property. This includes patents, trademarks, copyrights, industrial designs, and models. Companies operating in the industry must comply with intellectual property laws, as this protects their property and allows them to compete fairly in the market.

Another important aspect of the legal framework is environmental protection. Companies must comply with environmental laws, which establish standards and regulations for waste production and management, gas emissions, and water pollution. Companies that do not comply with these regulations may face fines and sanctions, as well as civil lawsuits.

Safety regulations

Safety in the workplace is a critical aspect in the industry. The safety and health of workers must be protected by law, and companies must comply with workplace safety regulations. This includes implementing safety programs, training workers in safety, and maintaining equipment and machinery in safe conditions.

In addition, companies must also comply with safety regulations regarding the transport and handling of hazardous materials. This includes the transportation of hazardous chemicals, handling of explosive materials, and disposal of toxic waste. Companies that do not comply with these regulations may face criminal and civil penalties.

Employment regulations

Companies operating in the industry must also comply with labor and employment laws. This includes the payment of fair wages, protection of workers' rights, and compliance with workplace safety and health regulations. Companies must also comply with laws against discrimination and harassment in the workplace.

In addition, companies must comply with employment regulations regarding the hiring and firing of workers. This includes compliance with child labor regulations and protection of the rights of migrant workers.

Trade regulations

Companies operating in the industry must also comply with trade regulations and antitrust laws. Trade laws establish rules for international trade, including standards and regulations for the import and export of goods. Companies must also comply with antitrust laws, which seek to prevent the formation of monopolies and promote fair competition in the market. Companies that do not comply with these regulations may face sanctions and fines.

Quality regulations

The quality of products and services is another important aspect in the industry. Companies must comply with quality regulations established by government authorities and corresponding regulatory bodies. This includes food safety regulations, which establish standards and regulations for the production and handling of food and beverages. Companies must also comply with quality regulations regarding the production and supply of pharmaceuticals and medical devices.

Tax regulations

Companies operating in the industry must also comply with tax and fiscal laws. This includes the payment of taxes, fees, and tariffs, as well as compliance with accounting and auditing regulations. Companies must submit accurate and transparent financial reports and comply with accounting regulations established by government authorities.

Property and zoning regulations

Companies operating in the industry must also comply with property and zoning regulations. Property laws establish rules for the ownership and use of land and buildings. Companies must comply with zoning regulations, which establish standards and regulations for the use of land and construction of buildings. Companies must obtain necessary permits before constructing or modifying a building, and comply with construction and safety regulations.

Conclusion

In summary, the industry is subject to a wide range of regulations and laws that seek to protect workers, the environment, and the general public, and promote fair competition in the market. Companies operating in the industry must comply with these regulations and laws in order to operate effectively and responsibly. It is important for companies to understand the legal and regulatory framework governing the industry in their country and ensure compliance with all applicable regulations and laws. In this way, companies can protect their intellectual property, keep their workers safe and healthy, produce high-quality products, and contribute positively to the economy and society as a whole.

SUSTAINABILITY AND SOCIAL RESPONSIBILITY IN INDUSTRIAL ENGINEERING

Sustainability and social responsibility are two key topics in modern industrial engineering. As natural resources become increasingly scarce and concerns over the environmental and social impact of human activities grow, industrial engineers are tasked with designing and operating production systems that are sustainable and socially responsible. In this chapter, we will explore the concepts of sustainability and social responsibility in industrial engineering, and discuss some of the tools and strategies that engineers can use to integrate these concepts into their work.

Sustainability in Industrial Engineering

Sustainability refers to the ability to meet present needs without compromising the ability of future generations to meet their own needs. In industrial engineering, sustainability involves designing and operating production systems in a way that efficiently utilizes resources and minimizes environmental impact. This involves adopting practices that reduce the use of non-renewable resources, decrease greenhouse gas emissions, reduce waste generation, and promote biodiversity conservation.

One of the most important tools that industrial engineers can use to improve the sustainability of production systems is life cycle assessment (LCA). LCA is a methodology that allows for the evaluation of the environmental impact of a product or system throughout its entire life cycle, from raw material extraction to final disposal. By using LCA, engineers can identify critical points in the life cycle of a product or system and take steps to reduce its environmental impact.

Another important tool is design for environment (DfE). DfE involves considering the environmental impacts of a product or system from the initial design stage and seeking opportunities to reduce its environmental impact throughout its entire life cycle. This may involve the use of more sustainable materials and technologies, reducing energy and resource use, and implementing recycling and waste management practices.

Social Responsibility in Industrial Engineering

Social responsibility refers to the obligation of companies and organizations to operate in an ethical and responsible manner with regard to society as a whole. In industrial engineering, social responsibility involves adopting practices that promote social justice, equal opportunities, and the well-being of the community at large.

One key area of social responsibility for industrial engineers is occupational safety and health management. Engineers must design and operate production systems that are safe and healthy for workers, and take steps to prevent accidents and injuries in the workplace. This may include implementing safety measures such as the use of personal protective equipment and safety training for workers.

Another key area of social responsibility is supply chain management. Engineers must work with suppliers and contractors to ensure that their practices are socially responsible and comply with ethical and legal standards. This may include promoting fair labor practices, eliminating child and forced labor, and adopting sustainable environmental management practices.

In addition, industrial engineers must also consider social responsibility in the design of products and systems. This involves considering how products and systems will affect end users and society as a whole, and seeking opportunities to improve quality of life and social equity. For example, engineers can design products that are accessible to people with disabilities or that promote gender equality.

Tools and Strategies for Sustainability and Social Responsibility in Industrial Engineering

To integrate sustainability and social responsibility into their work, industrial engineers can use a variety of tools and strategies. Some of the most common tools and strategies include:

Standards and norms: Engineers can use a variety of standards and norms to guide their practices and ensure sustainability and social responsibility in their projects. Examples include ISO 14001 (environmental management system), ISO 45001 (occupational health and safety management system), and SA8000 (social responsibility standard).

Certifications and labels: Engineers can work with certifications and labels to demonstrate that their products and systems meet environmental and social standards. Examples include the LEED (Leadership in Energy and Environmental Design) certification for green buildings, and fair trade labels for socially responsible manufacturing.

Social impact analysis: Engineers can use tools such as social impact analysis to evaluate the impact of their projects on society as a whole. This may involve assessing positive impacts such as job creation or access to basic services, as well as identifying negative impacts such as social exclusion or loss of cultural heritage.

Stakeholder engagement and dialogue: Engineers can involve stakeholders such as the local community, workers, and interest groups in the design and implementation of projects. This can help ensure that projects are socially responsible and meet the needs and concerns of all stakeholders.

Innovation and technology: Engineers can use innovation and technology to develop sustainable and socially responsible solutions. This may involve using clean and renewable technologies, implementing more efficient processes, and exploring new forms of design and production.

Conclusion

Sustainability and social responsibility are critical issues for modern industrial engineering. As the world faces increasingly complex environmental and social challenges, industrial engineers have a responsibility to design and operate production systems that are sustainable and socially responsible. This requires considering not only efficiency and economic profitability, but also the impact of their practices on people and the planet.

Industrial engineers have a wide variety of tools and strategies at their disposal to integrate sustainability and social responsibility into their work. These tools and strategies include standards and norms, certifications and labels, social impact analysis, stakeholder engagement and dialogue, and innovation and technology.

It is important for industrial engineers to work collaboratively with other professionals and stakeholders to ensure that their practices are truly sustainable and socially responsible. This may involve collaboration with environmental scientists, occupational health and safety experts, local community representatives, and other interest groups.

Sustainability and social responsibility in industrial engineering are complex and constantly evolving issues. Industrial engineers must stay informed about the latest developments in these areas and be willing to adapt their practices accordingly. Through a collaborative, sustainability- and socially-responsible approach, industrial engineers can play a crucial role in building a more just and sustainable world.

TRENDS AND FUTURE PERSPECTIVES IN INDUSTRY

Currently, the industry is undergoing constant change due to various factors such as technological innovation, globalization, digital transformation, sustainability, and circular economy, among others. These trends are impacting the way companies operate and how they relate to their customers and the environment. In this chapter, these trends will be analyzed and future perspectives of the industry will be explored.

Technological Innovation

Technological innovation is one of the main trends transforming the industry. The digitization of processes and automation of tasks are improving the efficiency and productivity of companies. In addition, the implementation of technologies such as artificial intelligence, the internet of things, virtual and augmented reality, and robotics are generating new business opportunities and improving the customer experience.

An example of how technological innovation is impacting the industry is the case of the automotive industry. The incorporation of technologies such as autonomous driving and connectivity are transforming the way people use vehicles and how they interact with them.

Globalization

Globalization is another trend that is transforming the industry. The opening of markets and internationalization of companies are generating new business opportunities and allowing access to new customers and suppliers. However, globalization also presents challenges such as global competition and the need to adapt to different cultures and regulations.

Globalization has allowed companies from various sectors to establish themselves in different countries and expand their presence in the global market. An example of this is the case of technology companies, which have been able to establish their presence in different countries and regions thanks to globalization and connectivity.

Digital transformation

Digital transformation is another trend that is impacting the industry. The digitization of processes and the implementation of technologies such as big data, artificial intelligence, and the cloud are allowing companies to improve efficiency and productivity, as well as offer new digital services and products.

Digital transformation is also enabling companies to improve the customer experience, thanks to the implementation of digital solutions that allow for more efficient communication and personalized interaction. An example of this is the case of e-commerce companies, which have improved the customer experience through the implementation of digital solutions that allow for a more personalized and efficient shopping experience.

Sustainability and circular economy

Sustainability and the circular economy are increasingly important trends in the industry. Sustainability involves the need to reduce the environmental impact of companies and their products, while the circular economy aims to reduce waste and resource use through the reuse and recycling of materials.

Sustainability and the circular economy are transforming the way companies operate and how they relate to the environment. An example of this is the case of fashion companies, which are implementing sustainability and circular economy strategies throughout the supply chain, from production to the sale and recycling of products. This involves the use of sustainable materials, waste reduction, and the implementation of recycling and reuse processes.

Future perspectives of the industry

As for the future perspectives of the industry, it is expected that the trends mentioned above will continue to transform the sector in the coming years. Below are some of the most relevant future perspectives that will be analyzed.

Artificial intelligence and automation

Artificial intelligence and automation are two trends that will continue to transform the industry in the future. It is expected that the implementation of these technologies will continue to improve the efficiency and productivity of companies, as well as enable the creation of new products and services.

In addition, artificial intelligence and automation also have the potential to generate new jobs and improve working conditions, by allowing workers to focus on more complex and creative tasks.

Circular economy and sustainability

The circular economy and sustainability will continue to be key trends in the industry in the future. It is expected that companies will continue to adopt sustainable practices and circular economy strategies, and that more and more consumers will demand sustainable products and services.

The implementation of circular economy and sustainability strategies can also be an opportunity for companies to differentiate themselves from the competition and improve their corporate reputation.

Emerging technologies

Emerging technologies such as virtual and augmented reality, blockchain, and quantum computing will also have an impact on the industry in the future. It is expected that these technologies will allow for the creation of new products and services and improve the customer experience.

In addition, the implementation of these technologies can be an opportunity for companies to innovate and differentiate themselves from the competition.

Digital transformation

Digital transformation will continue to be a relevant trend in the industry in the future. It is expected that companies will continue to digitize processes and adopt new technologies to improve efficiency and productivity.

In addition, digital transformation can also be an opportunity for companies to adapt to changes in consumer preferences and improve the customer experience.

Conclusions

In conclusion, the industry is constantly changing due to various trends such as technological innovation, globalization, digital transformation, sustainability, and the circular economy. These trends are transforming the way companies operate and how they relate to their customers and the environment.

In the future, it is expected that these trends will continue to transform the industry and new trends will emerge, such as artificial intelligence and automation, the circular economy and sustainability, emerging technologies, and digital transformation.

Therefore, it is important for companies to be prepared to adapt to these changes and take advantage of the opportunities that arise. Those that manage to adapt and adopt sustainable and innovative practices will be the ones that are most successful in the future of the industry.

ABOUT THE AUTHOR

Industrial and Systems Engineer

Master in Administration with Quality and Productivity

3 Certifications

2 Technical Studies

More than 20 training courses

Winner of the Singapore Cooperation Programme ITE

Instructor, engineer, content creator, and writer.
Discover the modern industry with the most controversial engineer.
engr's Workshop

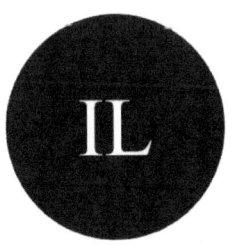

I. Laisequilla

Author / Engineer

RELATED BOOKS

the all about Lean Six Sigma

A complete and practical guide for those who wish to implement this methodology in their organization, with a detailed focus on the principles and theoretical foundations.

Available formats: physical, ebook, and audiobook

the all about Industrial Quality

FMEA, SPC, MSA, APQP, FMECA, Kaizen, Lean, ISO 9001, ISO 14001, ISO 45001, among others. Explained in an accessible and easy-to-understand manner.

Available formats: physical, ebook, and audiobook

the all about Industrial Methods

Lean Manufacturing, Six Sigma, Kaizen, TQM, Business Process Management. From classic methods to the most modern and emerging ones.

Available formats: physical, ebook, and audiobook

www.ingramcontent.com/pod-product-compliance
Lightning Source LLC
Chambersburg PA
CBHW070612220526
45467CB00003B/1395